PLASMA, CONSCIOUSNESS, AND THE CONTINUITY OF BEING

PLASMA, CONSCIOUSNESS, AND THE CONTINUITY OF BEING

DAKOTA FRANDSEN

First Printing, 2025

CONTENTS

Introduction

The enigmatic nature of consciousness has long been a profound source of human fascination and inquiry. Despite remarkable advancements in neuroscience, which have meticulously mapped the brain's intricate physical architecture, and in biochemistry, which has unveiled the molecular machinery underpinning thought processes, the subjective quality of awareness—often termed the "hard problem" of consciousness—continues to elude comprehensive scientific explanation. This essay ventures into an exploration of unconventional, yet thought-provoking frameworks that bridge the disciplines of physics, biology, and metaphysics, aiming to illuminate how consciousness might not merely reside within the confines of individual lives but could potentially persist, transfer, and even transcend them.

At the dynamic intersection of cutting-edge plasma physics, sophisticated cellular biology, profound quantum mechanics, and emerging consciousness studies, a compelling collection of theories is beginning to challenge traditional assumptions about the relationship between mind and matter. Accumulating evidence suggests that consciousness may possess qualities of fluidity, transferability, and endurance that extend beyond the boundaries conventionally recognized by mainstream scientific paradigms. This evidence spans a broad spectrum, from the inherent self-organizing characteristics observed in complex plasma structures and the compelling transgenerational inheritance of trauma via epigenetic mechanisms, to meticulously documented cases of children exhibiting verifiable memories of past lives and the intricate bioelectric fields that orchestrate the precise development of embryos.

This exploratory essay will systematically examine several interconnected phenomena that contribute to this broader understanding. First, it will delve into the distinctive properties of plasma and plasmoids, whose unique behaviors have prompted some researchers to propose novel plasma-based models of consciousness. Second, it will investigate the growing body of evidence supporting cellular and genetic memory, which suggests that information storage and recall may occur through mechanisms extending beyond the established neural networks of the brain. Third, the essay will explore the systematic investigation of reincarnation cases, presenting compelling accounts that defy conventional scientific explanations. Finally, it will analyze theoretical models that attempt to integrate these disparate and intriguing observations into coherent, overarching theories concerning the continuity and potential transcendence of consciousness.

Our approach to these complex and often controversial subjects is characterized by a spirit of genuine inquiry and open-mindedness, rather than the assertion of established dogma. It is important to acknowledge that some of the theories presented herein are inherently highly speculative and continue to be subjects of intense debate and controversy within the mainstream scientific community. Conversely, other concepts are firmly grounded in substantial empirical findings that unequivocally warrant serious and rigorous consideration. By examining these diverse perspectives collectively and holistically, we aim to uncover latent patterns, unexpected correlations, and novel possibilities that individual disciplines, often constrained by their specific methodological boundaries and paradigms, might otherwise inadvertently overlook.

Plasma, frequently referred to as the fourth state of matter, is not merely a theoretical construct but a ubiquitous and fundamental component of the universe, constituting an astounding approximately 99% of its visible mass. Unlike the more familiar states of matter—solids, liquids, or gases—plasma is composed of an ionized gas, a state where electrons have been stripped from their parent atoms, resulting in a dynamic and highly reactive mixture of charged particles. This unique and fundamental composition accounts for its pervasive presence in celestial bodies such as the incandescent interiors of stars, the vast expanses of interstellar clouds, and the energetic flow of the solar wind that permeates our solar system.

A paramount characteristic that critically differentiates plasma from other states of matter is its profound collective behavior. Rather than individual particles moving in a random and uncoordinated fashion, plasma particles engage in powerful, long-range electromagnetic interactions. These intricate and far-reaching interactions give rise to the formation of complex, often mesmerizing, self-organizing structures that exhibit properties remarkably akin to those observed in living organisms. Plasma possesses the intrinsic ability to establish distinct boundaries, generate internal electric fields that influence its own dynamics, react dynamically and adaptively to external disturbances, and maintain coherent, stable formations over immense cosmic distances and extended periods.

Hannes Alfvén, a distinguished Nobel Prize-winning Swedish physicist, conducted groundbreaking and extensive research in the field of magnetohydrodynamics. His seminal work definitively demonstrated plasma's remarkable capacity to conduct powerful electrical currents across vast cosmic expanses, leading to the formation of intricate filamentary structures that weave throughout the fabric of space. Alfvén's profound findings revealed that plasma is far from being a passive or inert medium; instead, it actively arranges itself into incredibly intricate and often beautiful geometric patterns. These formations include complex spirals, toroidal shapes, double layers (regions of strong electric fields), and vast interconnected networks, which, quite surprisingly, bear a striking resemblance to the hierarchical and interconnected structures found within biological neural networks. This observation opens intriguing avenues for speculation regarding the potential role of plasma in information processing and emergent consciousness.

The prevailing scientific consensus on consciousness, memory, and personal identity often struggles to fully account for a wealth of observed phenomena. Beyond the well-understood mechanisms of neural networks and genetic inheritance, there exists a compelling body of emerging research and long-standing empirical observations suggesting that information, and indeed fundamental aspects of consciousness itself, may persist and transfer through mechanisms far more intricate and expansive than currently recognized. This exploration ventures into these uncharted territories, examining the fascinating potential of plasma-based consciousness models, the distributed nature of biological memory systems, and compelling evidence for consciousness continuity, as powerfully illustrated by phenomena such as reincarnation and near-death experiences.

Essay

The pioneering work of Winston Bostick in the 1950s provided an early glimpse into the self-organizing capabilities of plasma. Bostick demonstrated the spontaneous formation of stable, particle-like plasmoids from ionized gas. These plasmoids exhibited remarkable properties, including self-organization, efficient energy exchange, and even a form of "memory" of past states, suggesting an intrinsic capacity for information retention. Building on this, V.N. Tsytovich and colleagues, in their seminal 2007 research, further posited that complex plasma structures could exhibit behaviors akin to living systems, showcasing replication and evolution in sophisticated simulations. These findings collectively open an intriguing avenue of inquiry: could consciousness itself involve plasma-like dynamics operating within the brain's intricate bioelectric fields?

The brain, at its core, functions as an electrical organ, ceaselessly generating measurable electromagnetic fields that extend far beyond the confines of individual neurons. Theories like Johnjoe McFadden's Conscious Electromagnetic Information (CEMI) Field Theory propose that consciousness does not solely emerge from the discrete firing of neurons, but rather from the integrated information within these overarching electromagnetic fields. The brain's prodigious electrical activity, meticulously driven by the precise movement of charged ions, could exhibit collective behaviors strongly analogous to plasma, including complex self-organization and the storage of information within dynamic spatial patterns. Even the enigmatic realm of quantum coherence, as articulated in the Penrose-Hameroff Orchestrated Objective Reduction (Orch OR) Theory, suggests that quantum states within the brain's microtubules might play a crucial role in facilitating consciousness. This concept has gained further traction with recent discoveries demonstrating quantum phenomena in various biological systems, such as the highly efficient process of photosynthesis.

Several compelling analogies bridge the seemingly disparate fields of plasma physics and consciousness:

- **Self-Organization:** Both plasmoids and consciousness profoundly display the emergence of highly complex, coherent structures and behaviors from the coordinated interaction of individual, simpler components. This emergent property is a hallmark of both systems.
- **Field-Based Information Processing:** Plasma behavior is predominantly governed by intricate electromagnetic field interactions. In a similar vein, consciousness might fundamentally involve the integration and processing of information at the level of electromagnetic fields, rather than solely through discrete neural connections.
- **Non-Locality:** Plasma structures have been observed to exhibit correlated behavior across significant distances, hinting at a form of non-local communication or influence. This parallels the profound "binding problem" in consciousness, where diverse and

spatially distributed brain regions somehow integrate seamlessly into unified, coherent perceptions.

- **Memory and Persistence:** Plasmoids demonstrate a remarkable ability to maintain stable configurations and, as Bostick noted, "remember" past states. Consciousness, too, exhibits profound continuity and enduring memory, even in the face of the constant and dynamic changes occurring within its neural substrate.
- **Resonance and Coherence:** Plasma systems possess specific resonant frequencies and can enter highly coherent states. This resonates strongly with the observed brain rhythms (e.g., gamma waves) that are consistently correlated with states of conscious awareness and cognitive function.

These powerful analogies strongly suggest that plasma physics offers invaluable theoretical and analytical tools for modeling the intricate nature of consciousness, potentially illuminating aspects of neural information processing that extend far beyond conventional neuron-centric perspectives.

Beyond the Brain: Distributed Information and the Enigma of Biological Memory

The traditional understanding of memory predominantly localizes its storage within the brain's intricate synaptic connections. However, a growing body of compelling evidence increasingly points towards a far more distributed and multifaceted system of information storage throughout the entire human body:

- **Cellular Memory:** Remarkable anecdotal and some systematic reports from organ transplant recipients occasionally describe inexplicable changes in preferences, behaviors, or even the emergence of memories that are remarkably consistent with those of their donors. These intriguing cases strongly suggest that cells outside the nervous system might possess the capacity to store and retrieve experiential information.
- **Epigenetic Memory:** Environmental factors possess the profound ability to modify gene expression without altering the underlying DNA sequence. Crucially, these

epigenetic changes can be inherited across generations. Studies have demonstrated the transgenerational inheritance of learned fear responses in mice and the presence of trauma-related methylation patterns in humans, indicating a biological memory that transcends individual experience.

- **Bioelectric Codes:** Michael Levin's groundbreaking research highlights the critical role of bioelectric fields, which are present across all tissues, in guiding development and regeneration. These fields store vast amounts of large-scale anatomical information that astonishingly persists across multiple cell generations. These complex voltage patterns within cells and tissues could, in theory, encode and store experiential information, forming a dynamic bioelectric memory.
- **Morphic Resonance:** Rupert Sheldrake's highly controversial, yet thought-provoking, hypothesis of morphic resonance proposes that organisms inherit patterns of form and behavior not solely through genetics, but also from previous members of

their species via "morphic fields." This radical concept suggests a form of biological memory that operates beyond the confines of individual organisms, tapping into a collective, species-specific memory.

Consciousness Transference Through Biology: Unseen Pathways

The intricate networks of cellular communication within the body, encompassing chemical signals, finely tuned electromagnetic fields, and even the subtle influence of quantum coherence, strongly hint at the existence of non-conventional mechanisms for information transfer. The human microbiome, inherited from mothers, represents another profound layer of biological information transfer, playing a crucial role in influencing offspring development and health. Recent theoretical work has even ventured into exploring DNA's potential as a quantum antenna, suggesting its capacity to process information in ways that extend far beyond its conventionally understood chemical role.

Empirical Investigations into Reincarnation: Challenging Materialist Assumptions

Systematic and rigorous research into claims of reincarnation provides some of the most compelling, albeit controversial, evidence for the continuity of consciousness beyond physical death:

- **Ian Stevenson's Studies:** Dr. Ian Stevenson dedicated his career to meticulously documenting over 2,500 cases of children who reported detailed memories of past lives. These accounts often included verifiable specifics about deceased individuals whom the children could not have known through conventional means. His extensive work includes striking cases where birthmarks or congenital defects on the child precisely matched fatal wounds sustained by the alleged previous personality.
- **Jim Tucker's Research:** As Stevenson's successor, Dr. Jim Tucker continues this rigorous research tradition, expanding the focus to include cases predominantly from the United States. Notable examples include

James Leininger, a young child who exhibited remarkably detailed knowledge of WWII aircraft and specific memories consistent with those of a deceased pilot. Tucker's work also delves into the fascinating and often profound memories of the "intermission period" between lives, as reported by some children.

While skeptics naturally raise valid concerns regarding potential explanations such as fraud, cryptomnesia (unconscious plagiarism), coincidence, selective reporting, leading questions, and cultural conditioning, researchers in this field actively address these challenges through meticulous methodologies, including blind verification of details, rigorous statistical analysis, and cross-cultural studies to identify universal patterns.

Theoretical Frameworks for Consciousness Persistence: An Integrated View

Several emerging theoretical frameworks attempt to integrate these diverse and often perplexing phenomena into a more cohesive understanding of consciousness:

- **Information as Fundamental:** Theories positing information as a fundamental constituent of reality suggest that consciousness is essentially an intricate information pattern that can manifest and exist across multiple physical substrates. David Bohm's concept of the implicate order, for instance, suggests that consciousness might be localized expressions of a more fundamental, unified field of information that underlies all reality.
- **Quantum Information and Non-Locality:** Quantum mechanics, with its profound implications of entanglement and non-local correlations, introduces the possibility of information transfer that transcends classical physics. If consciousness involves quantum information processing, its persistence and transfer might similarly transcend the limitations of the classical physical world.

- **Bioelectric Field Templates:** As suggested by Levin's research, bioelectric fields could serve as dynamic templates that store experiential information. This information might then potentially persist after the death of the organism or influence the development and experiences of subsequent biological systems.
- **Morphic Resonance as an Information Channel:** Sheldrake's theory posits that strong, recurring patterns of thought and experience create morphic fields. These fields, through a process of resonance, could potentially influence future consciousnesses, acting as a non-physical information channel, much like a radio tuning to a specific frequency.

An integrated and multi-layered model of consciousness could envision it as a complex interplay of various levels: quantum information processing, dynamic bioelectric fields, resonant morphic fields, and plasma dynamics, all interfacing with intricate biological memory systems. Within such a framework, compelling reincarnation cases might be understood as a unique alignment or resonance of these various levels, where a strong morphic field and specific, resonant biological patterns facilitate the remarkable transfer of information from one life to the next. This holistic perspective offers a rich and expansive lens through which to explore the enduring mysteries of consciousness.

Near-Death Experiences (NDEs) and Consciousness Continuity

NDEs offer another line of evidence for consciousness beyond brain function. Cardiac arrest survivors report vivid, coherent experiences during periods of no electrical brain activity, challenging materialist models. Studies by Pim van Lommel and Sam Parnia have documented verifiable observations made during NDEs when consciousness should have been absent. The consistent phenomenology of NDEs across cultures further supports their validity.

These experiences suggest the brain might act as a receiver or filter for consciousness, rather than its sole generator. If consciousness involves field phenomena (electromagnetic, bioelectric, quantum, or morphic), these might maintain coherence even when neural activity is disrupted.

The Continuity Question and Implications

The combined evidence from NDEs, reincarnation, and cellular memory points to a profound question: is consciousness a continuous process extending beyond individual lifetimes? While challenges remain (e.g., rarity of memories, alternative explanations, lack of a known physical mechanism for full memory transfer), a middle ground suggests that certain patterns (emotional imprints, trauma, dominant traits) might persist through quantum, bioelectric, or morphic channels, influencing new developing consciousnesses without requiring full continuity.

Practical Applications and Future Directions

These frameworks have potential medical and therapeutic applications:

- **Consciousness Disorders:** New interventions for coma or vegetative states could target bioelectric fields or electromagnetic resonances.
- **Trauma and PTSD:** Addressing inherited epigenetic patterns could enhance trauma therapy.
- **Organ Transplantation:** Understanding cellular memory could inform recipient support and protocols.
- **Reincarnation Cases:** Therapies for children with distressing past-life memories could acknowledge their phenomenological reality.

Future research paths include detailed bioelectric mapping, quantum coherence detection in neural tissue, rigorous replication of reincarnation studies, investigation of electromagnetic correlates of consciousness, and exploration of epigenetic encoding of specific information.

Challenges and Philosophical Considerations

Significant challenges remain, including scientific and methodological issues (e.g., difficulty demonstrating plasma-like behavior in the brain, maintaining quantum coherence, lack of empirical support for specific memory transfer via cellular mechanisms, and alternative explanations for reincarnation). The replication crisis in parapsychological research also highlights the need for rigorous methodology.

Integrating these diverse frameworks into a coherent theory is a formidable task, requiring explanations for scaling quantum processes to bioelectric fields, interfacing bioelectric with morphic fields, and explaining the rarity of reincarnation memories.

Philosophically, the "hard problem" of subjective experience remains unsolved. These frameworks describe how consciousness might operate, but not why it arises. Panpsychism or idealism, which view consciousness as fundamental, offer alternative perspectives.

Ethically, these ideas raise profound questions about personal identity, responsibility (e.g., inherited trauma), medical ethics (end-of-life decisions, organ donation), enhancement ethics, and animal consciousness. Culturally and religiously, these phenomena intersect with ancient traditions, informing dialogue between science and spirituality.

Toward an Integrated Understanding

The exploration of these diverse fields suggests that consciousness, life, and information are interwoven in subtle and complex ways. Key findings indicate:

1. Biology stores and transfers information beyond genetics (epigenetics, immune memory, microbiome, bioelectric fields).

2. Consciousness correlates with electromagnetic field phenomena, and plasma physics offers useful models.
3. Some children report verifiable memories of previous lives, warranting scientific attention.
4. Consciousness can apparently persist during periods of severely compromised brain function (NDEs).
5. Quantum coherence operates in biological systems, making quantum consciousness theories less implausible.

These findings do not definitively prove plasma-based consciousness or reincarnation but suggest that the boundaries between organisms, matter, and information are more permeable than previously thought. Consciousness might involve multiple levels of organization, each with its dynamics for information storage and transfer.

Perhaps consciousness is a dynamic pattern that can instantiate itself in various substrates, like a wave moving through different media. An individual consciousness could be analogous to a plasmoid within a larger, fundamental consciousness field. This speculative synthesis invites continued research and rigorous investigation, promising to transform our understanding of what it means to be a conscious being in a vast and enigmatic universe.

REFERENCES AND BIBLIOGRAPHY

I. Plasma Physics and Plasmoids

1. Alfvén, H. (1981). *Cosmic Plasma*. D. Reidel Publishing Company.
2. Bostick, W. H. (1956). "Experimental Study of Ionized Matter Projected across a Magnetic Field." *Physical Review*, 104(2), 292-299.
3. Chen, F. F. (2016). *Introduction to Plasma Physics and Controlled Fusion* (3rd ed.). Springer.
4. Lerner, E. J. (1991). *The Big Bang Never Happened*. Times Books.

5. Peratt, A. L. (1992). *Physics of the Plasma Universe*. Springer-Verlag.
6. Tsytovich, V. N., et al. (2007). "From plasma crystals and helical structures towards inorganic living matter." *New Journal of Physics*, 9(8), 263.

II. Consciousness and Physics

1. Chalmers, D. J. (1995). "Facing up to the problem of consciousness." *Journal of Consciousness Studies*, 2(3), 200-219.
2. Hameroff, S., & Penrose, R. (1996). "Orchestrated reduction of quantum coherence in brain microtubules: A model for consciousness." *Mathematics and Computers in Simulation*, 40(3-4), 453-480.
3. Koch, C. (2019). *The Feeling of Life Itself: Why Consciousness Is Widespread but Can't Be Computed*. MIT Press.
4. McFadden, J. (2002). "The conscious electromagnetic information (cemi) field theory: The hard problem made easy?" *Journal of Consciousness Studies*, 9(8), 45-60.

5. Penrose, R. (1994). *Shadows of the Mind: A Search for the Missing Science of Consciousness*. Oxford University Press.
6. Tononi, G. (2004). "An information integration theory of consciousness." *BMC Neuroscience*, 5(1), 42.

III. Cellular and Genetic Memory

1. Bale, T. L. (2015). "Epigenetic and transgenerational reprogramming of brain development." *Nature Reviews Neuroscience*, 16(6), 332-344.
2. Dias, B. G., & Ressler, K. J. (2014). "Parental olfactory experience influences behavior and neural structure in subsequent generations." *Nature Neuroscience*, 17(1), 89-96.
3. Lipton, B. H. (2005). *The Biology of Belief: Unleashing the Power of Consciousness, Matter and Miracles*. Mountain of Love Productions.
4. Pearsall, P., Schwartz, G. E., & Russek, L. G. (2002). "Changes in heart transplant

recipients that parallel the personalities of their donors." *Journal of Near-Death Studies*, 20(3), 191-206.

5. Pembrey, M. E., et al. (2006). "Sex-specific, male-line transgenerational responses in humans." *European Journal of Human Genetics*, 14(2), 159-166.
6. Sheldrake, R. (2009). *Morphic Resonance: The Nature of Formative Causation*. Park Street Press.
7. Yehuda, R., et al. (2016). "Holocaust exposure induced intergenerational effects on FKBP5 methylation." *Biological Psychiatry*, 80(5), 372-380.

IV. Reincarnation Research

1. Matlock, J. G. (1990). "Past life memory case studies." In S. Krippner (Ed.), *Advances in Parapsychological Research* (Vol. 6, pp. 184-267). McFarland.
2. Mills, A., Haraldsson, E., & Keil, H. H. J. (1994). "Replication studies of cases suggestive of reincarnation by three indepen-

dent investigators." *Journal of the American Society for Psychical Research*, 88(3), 207-219.

3. Stevenson, I. (1997). *Reincarnation and Biology: A Contribution to the Etiology of Birthmarks and Birth Defects* (2 volumes). Praeger.
4. Stevenson, I. (2001). *Children Who Remember Previous Lives: A Question of Reincarnation* (revised edition). McFarland & Company.
5. Tucker, J. B. (2005). *Life Before Life: A Scientific Investigation of Children's Memories of Previous Lives*. St. Martin's Press.
6. Tucker, J. B. (2013). *Return to Life: Extraordinary Cases of Children Who Remember Past Lives*. St. Martin's Press.

V. Near-Death Experiences and Consciousness Continuity

1. Alexander, E. (2012). *Proof of Heaven: A Neurosurgeon's Journey into the Afterlife*. Simon & Schuster.

2. Greyson, B. (2010). "Implications of near-death experiences for a postmaterialist psychology." *Psychology of Religion and Spirituality*, 2(1), 37-45.
3. Parnia, S., et al. (2014). "AWARE—AWAreness during REsuscitation—A prospective study." *Resuscitation*, 85(12), 1799-1805.
4. van Lommel, P., et al. (2001). "Near-death experience in survivors of cardiac arrest: A prospective study in the Netherlands." *The Lancet*, 358(9298), 2039-2045.

VI. Quantum Biology

1. Ball, P. (2011). "Physics of life: The dawn of quantum biology." *Nature*, 474(7351), 272-274.
2. Engel, G. S., et al. (2007). "Evidence for wavelike energy transfer through quantum coherence in photosynthetic systems." *Nature*, 446(7137), 782-786.

3. Lambert, N., et al. (2013). "Quantum biology." *Nature Physics*, 9(1), 10-18.
4. McFadden, J., & Al-Khalili, J. (2014). *Life on the Edge: The Coming of Age of Quantum Biology*. Crown Publishers.

VII. Neuroscience and Memory

1. Kandel, E. R. (2006). *In Search of Memory: The Emergence of a New Science of Mind*. W.W. Norton & Company.
2. Mayford, M., Siegelbaum, S. A., & Kandel, E. R. (2012). "Synapses and memory storage." *Cold Spring Harbor Perspectives in Biology*, 4(6), a005751.
3. Nader, K., Schafe, G. E., & Le Doux, J. E. (2000). "Fear memories require protein synthesis in the amygdala for reconsolidation after retrieval." *Nature*, 406(6797), 722-726.

VIII. Bioelectricity and Morphogenesis

1. Levin, M. (2014). "Molecular bioelectricity: How endogenous voltage potentials control cell behavior and instruct pattern regulation in vivo." *Molecular Biology of the Cell*, 25(24), 3835-3850.
2. Levin, M., & Martyniuk, C. J. (2018). "The bioelectric code: An ancient computational medium for dynamic control of growth and form." *BioSystems*, 164, 76-93.
3. McCaig, C. D., Rajnicek, A. M., Song, B., & Zhao, M. (2005). "Controlling cell behavior electrically: Current views and future potential." *Physiological Reviews*, 85(3), 943-978.

IX. Philosophical and Theoretical Works

1. Bohm, D. (1980). *Wholeness and the Implicate Order*. Routledge.
2. Chalmers, D. J. (1996). *The Conscious Mind: In Search of a Fundamental Theory*. Oxford University Press.

3. Nagel, T. (1974). "What is it like to be a bat?" *The Philosophical Review*, 83(4), 435-450.
4. Radin, D. (2006). *Entangled Minds: Extrasensory Experiences in a Quantum Reality*. Paraview Pocket Books.
5. Sheldrake, R. (1981). *A New Science of Life: The Hypothesis of Morphic Resonance*. Blond & Briggs.

X. Cross-Cultural and Historical Perspectives

1. Bowman, C. (1997). *Children's Past Lives: How Past Life Memories Affect Your Child*. Bantam Books.
2. Head, J., & Cranston, S. L. (1977). *Reincarnation: The Phoenix Fire Mystery*. Julian Press/Crown Publishers.
3. Shroder, T. (1999). *Old Souls: The Scientific Evidence for Past Lives*. Simon & Schuster.
4. Weiss, B. L. (1988). *Many Lives, Many Masters*. Simon & Schuster.

XI. Recent Interdisciplinary Research

1. Beauregard, M., & O'Leary, D. (2007). *The Spiritual Brain: A Neuroscientist's Case for the Existence of the Soul*. HarperOne.
2. Cardeña, E., Lynn, S. J., & Krippner, S. (Eds.). (2014). *Varieties of Anomalous Experience: Examining the Scientific Evidence* (2nd ed.). American Psychological Association.
3. Kelly, E. F., et al. (2007). *Irreducible Mind: Toward a Psychology for the 21st Century*. Rowman & Littlefield Publishers.
4. Radin, D., Michel, L., Galdamez, K., et al. (2012). "Consciousness and the double-slit interference pattern: Six experiments." *Physics Essays*, 25(2), 157-171.

Note: This bibliography encompasses a broad spectrum of viewpoints, ranging from established physics and neuroscience to emerging research in consciousness studies and reincarnation. The inclusion of a source signifies its contribution to the ongoing sci-

entific and philosophical discourse on these subjects, rather than an endorsement of all its claims.

www.ingramcontent.com/pod-product-compliance
Lightning Source LLC
LaVergne TN
LVHW022031200926
848481LV00007B/2640